Chemistry Case Studies for Allied Health

Colleen Kelley
Pima Community College

Wendy Weeks
Pima Community College

JOHN WILEY & SONS, INC.

Cover Photo: ©Cris Benton, Kite Aerial Photography

To order books or for customer service please, call 1-800-CALL WILEY (225-5945).

ISBN-13 978- 0-470-03976-2
ISBN-10 0-470-03976-0
Printed in the United States of America

10 9 8 7 6 5 4 3 2 1

Printed and bound by Bind Rite Graphics

Table of Contents

Introduction

Welcome to the world of learning through case studies. These cases provide an excellent opportunity for you to apply what you're learning in class and to assimilate this knowledge with a pseudo real-world scenario (called a case study) to solve or diagnose a problem. When approaching these case studies, it's important to keep in mind that while each case study is placed within a chapter with the most relevance, some of the tools needed to address the case study will also be found in preceding chapters. Learning in this way is dimensional - forward, backward, and circular. It's also a way to discover chemistry by using the scientific method. Each case study gives you the opportunity to apply the scientific method to solve a chemistry-related problem.

At first this task may seem daunting, but with a few guidelines and helpful hints, you'll be able to assume the role of diagnostician in no time. While we'll provide some specific guidance for each case study, there are a few approaches that are applicable to all of the cases.

Read through each case study at least two times.

Make a list of what you know (what's presented in the case study) and what you need to know. To illustrate this, read the following excerpt from a case study:

> *A 4-year old boy is brought by his mother into the emergency room on a Friday night. The boy is complaining of severe tooth pain. The attending physician takes a look in his mouth and notices substantial swelling in his gums and cheeks and the appearance of an abscessed tooth. The physician asks his mother some medical questions and discovers that he's a healthy 4-year old with no history of allergies. The mother has been giving her son children's ibuprofen in a suspension formula and has also been putting OraGel® on the painful area. The physician prescribes an antibiotic to treat the infection (the abscessed tooth) and recommends to the mother to continue giving her son ibuprofen to help with the pain and swelling.*

What do you know?

- 4-year old boy
- abscessed tooth
- no allergies
- boy has been taking ibuprofen and OraGel®
- physician prescribes antibiotic

What else might you need to know in order to address the pertinent chemistry?

- weight of the boy
- name/structure of the antibiotic prescribed
- structure of ibuprofen and OraGel®
- other?

From your list of what you need to know, write down possible reference materials for each item. While most of your research to answer these questions can be done on the internet, make sure you are using reputable sources.

One piece of information that you will frequently need to find is the molecular structure of a medication. There are many resources available and we can suggest using www.chemfinder.com or the manufacturer's website as a start.

Use your reference materials from the step above to answer your questions. If you are not able to answer your questions from these references, go back and find more.

If there are parts of the case study that can't be answered from a reference source, like the weight of the boy, then it's fine to make assumptions. Just make sure you clearly state your assumption as part of your diagnosis. In the example stated above, the weight of the 4 year old boy might make a difference in your answer, but there's no way to find out his actual weight from a reference. So, make the assumption that he's 45 pounds. Then, based on that assumption, you can make your recommendation. Of course, in the real world, you wouldn't need to make that assumption - you could get his actual weight.

Ask yourself again and again, "What is the pertinent ***chemistry*** in this case study and how will that help me to address the situation?" Each case study was designed to bring forward chemical concepts (general, organic, and/or biochemical) that will enable you to address the scenario.

Speaking of the real world... these case studies were designed to enhance your understanding of the role of chemistry in a particular vignette. With that in mind, it's feasible that many of the cases, while based on real situations, may not be "possible". So, refrain from straying from the case study or from approaching the case study as "that would never happen, so I'm not going to answer it". Remember, it's a hypothetical situation (based on real experiences) designed to enhance your understanding of chemistry and to make you a better diagnostician. In order to bring forward the chemical nature of the problem, some of the clinical aspects were exaggerated or modified.

Speaking of straying... all too often students, especially while solving one of his or her first case studies, go off on tangents that are interesting, but not related to the pertinent chemistry of the case study. Remember to focus on solving the chemistry of the case study. This will most likely necessitate a chemical structure of the molecules involved and perhaps some reactions of these molecules. In the example we've been using, the structure of ibuprofen will provide you with more chemical insight than will a discussion of the pros and cons of prescription antibiotics.

Finally, remember that there are many right answers. If you provide a correct chemical analysis and can use this to support your diagnosis, then it's a reasonable recommendation. On the other hand, an incorrect chemical analysis will not allow you to make a reasonable recommendation.

We're excited for you to experience learning chemistry in this realm. Our students have commented to us that learning through case studies was one of the most challenging academic experiences they've ever had and, at the same time, one of the most satisfying and enriching experiences.

Enjoy!

Chapter 1: Science and Measurement

Related Learning Objectives

Explain the following terms:
"scientific method"
"law"
"theory"
"hypothesis"
"experiment"

Convert from one unit of measurement into another.

Express values using scientific notation and metric prefixes.

CASE 1: Uncovering the Source of a Coma

Purpose: The purpose of this case study is to apply the scientific method by making empirical observations, formulating a hypothesis, testing the hypothesis, revising the hypothesis, and eventually asserting a theory.

> LG, is a 35 yo, Hispanic female who presents to the ER with stiff muscles, is sweaty and shaky, and she is short of breath. LG had experienced these symptoms before and the DX was hypokalemia. LG has a HX of spina bifida since birth, asthma, a latex allergy, has an illeostomy, and leg weakness which necessitates the use of a walker.

Terms to define

yo ______________________________

DX ______________________________

Hypokalelmia ______________________________

HX ______________________________

spina bifida ______________________________

illeostomy ______________________________

c/o ______________________________

bilaterally ______________________________

TX ______________________________

IV ______________________________

r/o ______________________________

via ______________________________

nebulizer ______________________________

pt ______________________________

K vs. K^+ ______________________________

Answer these questions before reading further.

1. What additional information do you want to know from this patient?

   ____

2. What medical assessments do you want to make?

   ____

LG's blood tests show low potassium levels; all other results are normal.

LG c/o trouble breathing due to her asthma but medical personnel believe she's panicking due to her symptoms, because breath sounds are normal and clear bilaterally. LG's continued complaints about her inability to breathe and her asthma prompt a nebulizer TX. IV infusion of potassium is also administered at this time.

One hour later LG goes into a coma.

3. What are possible reasons for her coma? Be sure and check her medical HX.

____

____

r/o Spina bifida and illeostomy.

4. What remaining DX do we still need to check?

____

____

5. Could any of these be a cause of a coma?

____

____

While in a coma, over the next two days, K^+ levels are brought back up to normal range and nebulizer breathing TX is continued via breathing mask.

r/o Coma due to asthma.

Coma is narrowed down to two possibilities; extended low K^+ level damage and latex allergy.

6. Keeping her HX in mind, what ailments are still suspect as to the reason for the coma?

____

____

7. Now what questions should we ask?

____

____

8. What organs or physiological process can be affected by potassium levels?

____

____

Further information: All medical personnel having contact with LG did not use latex gloves. Functions of organs affected by low potassium levels are normal. Nebulizer breathing TX continues.

A large sign saying 'Allergic to latex' has been posted on the door to LG's room and around her bed since she was put in the room. Being scrupulous about the latex because all indicators are pointing to that as the culprit, they assess every possible item that contains latex.

9. What other products contain latex?

____

____

Upon research and questioning, it was confirmed no latex had gotten near the pt via food, nebulizer, O_2 mask, gloves, tape, or anything from the hospital's end.

The nebulizer treatment was stopped and twenty-four hours later LG awakens from her coma.

10. What questions do you have for LG?

____

____

____

The culprit <u>is</u> the latex that induced LG's coma.

11. How was LG exposed to latex?

____

____

____

12. How did the latex create this physiological response?

____

____

____

CASE 2: How Much Is Too Much?

Purpose: The purpose of this case study is to calculate the daily total dose of an active ingredient in multiple medications and to determine if this cumulative dose could be responsible for the symptoms described.

LT and her roommate ER are two college juniors and have just returned from Spring break.

"Spring break was great," remarks LT. "But enough parties, it's time to hit the books now. I have three huge mid-term exams next week."

"My allergies are killing me," continues LT. "Do you have anything for them?"

"Sure," says ER. "I have some Benadryl® Maximum Strength Severe Allergy and Sinus."

"Sounds just like what I need," says LT. "It says here that I shouldn't take more than 8 of these a day. OK, I'll take 8 per day - that should take care of these allergies."

Five days later, ER asks LT about her allergies. "They're fine," remarks LT. "I'll keep taking the Benadryl® just to be sure."

"Look at your desk – it looks like a pharmacy," remarks ER.

"I know. Besides the Benadryl, I've been taking Maximum Strength Midol® PMS Caplets for you-know-what and, since all of these midterms are giving me a headache, I've had to take these Excedrin® Migraine Geltabs for a few days as well. At least I know enough not to exceed the maximum daily dosage of each," says LT.

1. What are the active ingredients in each of these medications?

____

____

2. How much of each active ingredient is contained in each tablet?

____

____

3. How much of each active ingredient is LT taking per day?

____

____

A few days later ER walks into her room and discovers LT in bed. "Hey, what's the matter, don't you have exams today?" asks ER.

"I don't feel well. I've been throwing up and have had diarrhea all day," says LT.

"Hey, you look yellow – like really yellow!" screams ER. "I'm going to take you to the emergency room now!"

When they arrive at the emergency room, the physician examines LT and asks about her medical history to include recent medications. She proclaims to have followed the directions on taking the maximum dosages per day of Benadryl®, Midol®, and Excedrin®. The physician is very concerned and immediately orders some laboratory work. He also notes she weighs 125 pounds.

4. Why is the physician concerned?

____

____

____

____

____

5. What do you suspect has occurred in this case?

____

____

____

____

____

6. What do LT's symptoms indicate?

____

____

____

____

____

Chapter 2: Atoms and Elements

Related Learning Objectives

Define the terms:
"element"
"atomic symbol"

Define the term "nuclear radiation" and describe the four common types of radiation emitted by radioisotopes. Explain how exposure to radiation can be controlled.

CASE 3: Chelated Calcium

Purpose: The purpose of this case study is to become familiar with elements as minerals that are essential to the nutrition of humans and the consequences of inadequate consumption or bioavailability.

RR, a 45-year-old woman, has a familial history of osteoporosis. Aware that she was a likely candidate for this disease, her physician orders a DEXA bone densitometry. Her Z score was -2.4. RR was referred to a nutritionist. During the evaluation process, the nutritionist asked RR to provide information about her typical dietary habits. RR provided the following information that caused the nutritionist concern.

Six Ultra (Maximum Strength Tums®) first thing in the morning.

Cooked spinach

Chocolate milk

Cooked beet greens

Stewed rhubarb

Four cans of 12 ounce cola per day

The nutritionist calculated RR's calcium intake, in the form of DRI's; it appeared RR was consuming an appropriate amount of calcium.

1. Explain to RR why she was sent to a nutritionist.

____

____

2. If RR appeared to be consuming an appropriate amount of calcium, why did the supplements and foods listed above cause the nutritionist concern?

____

____

____

3. Since we know that patient compliance is a serious problem, especially when it comes to dietary recommendations, what dietary modifications could you suggest to RR that would not be a large deviation from her typical nutritional habits?

____

____

CASE 4: The Discovery of Guacamolium (Gu)

Purpose: The purpose of this case study is to evaluate the efficacy of an alpha emitter for use in treating a tumor.

A medical researcher proposes using a newly found element, guacamolium (Gu), an alpha emitter, as a source of radiation therapy to destroy deep-seated, cancerous tumors. Since you are his direct supervisor, you must decide whether to fire him or submit his name for the Nobel Prize in medicine.

1. What is your decision? Support your decision with appropriate assumptions.

CASE 5: Running from Radiation

Purpose: The purposes of this case study are 1) to clarify public misconceptions about emissions from nuclear power plants, and 2) to decide whether or not the emissions from Three Mile Island could have been responsible for certain health problems.

On March 28, 1979, 7th grade students in Mechanicsburg, PA were told school was going to be closing and buses would be coming to take them home. Their teacher explained to the inquiring students that the nuclear power plant at the nearby Three Mile Island was "about to blow up." The students, being 7th graders, rejoiced and immediately opened the windows in the classroom to "see the explosion and the radioactivity". The teacher shrieked and told them to close the windows immediately so that "the radiation would not get into the school". The students were told to use their coats for shelter as they walked to the bus to "shield them from the radioactivity". Finally, they were instructed to "run home to avoid exposure to the radioactivity."

1. Are the recommendations made to the students regarding protecting themselves from "radioactivity" sound? Why or why not?

Weeks after the incident at Three Mile Island nuclear power plant, over 2000 personal injury claims were filed by attorneys. The plaintiffs were claiming a variety of health-injuries caused by gamma radiation. Fifteen years later, Judge Sylvia Rambo dismissed these cases and granted summary judgment in favor of the defendants.

2. Do you agree or disagree with Judge Rambo's verdict? Why or why not?

3. Why did it take fifteen years to render a verdict?

Chapter 3: Compounds

Related Learning Objectives

Define the term ion and explain how the electron dot structure of a representative element atom (groups IA - VIIIA) can be used to predict the charge on the monoatomic ion that it forms.

Describe the naming of monoatomic and polyatomic cations and anions.

Explain the difference between an ionic bond and a covalent bond.

Name and write the formulas of simple ionic compounds and binary molecules.

Define the terms formula weight and molecular weight and use these weights in unit conversions involving moles and mass.

CASE 6: Swimming in Chromium

Purpose: The purpose of this case study is to examine the different ions and ionic compounds formed by chromium and to gain an understanding of the toxicity of hexavalent chromium.

> You have just finished watching the movie "Erin Brokovich" and while you're still wondering how Julia Roberts won an Academy Award for this 2000 film, you're also wondering about the significance of the words "HEXAVALENT CHROMIUM" that were flashed across the screen while she was copying documents. After some research, you discover the events in this film recount actual events and resulted in the lawsuit "Anderson v Pacific Gas and Electric."

1. What is hexavalent chromium and why was it significant in this lawsuit?

____

____

2. Where did this hexavalent chromium come from? What's its formula and formula weight?

____

____

As you research more about the lawsuit, you come across a flyer distributed by Pacific Gas and Electric (PG&E):

> "Chromium occurs in two forms. The form that is present in groundwater can cause health effects in high doses. The cleanup program, however, will result in chromium levels that meet the very conservative drinking water standards set by the EPA. In addition, the form of chromium that will be left on soils after irrigation is nontoxic. In fact, chromium in this form is a naturally occurring metal that is an essential ingredient in the human diet, one that is often included in multiple vitamin/mineral supplements." [Quote taken from the Anderson v PG&E case - Superior Court for the County of San Bernardino, Barstow Division, File BCV 00300].

3. What are the two forms of chromium mentioned in the flyer?

____

4. What are the EPA standards for chromium in drinking water? Which kind(s) of chromium are included in this standard?

____

____

____

5. Why is one form of chromium toxic and one form considered essential?

____

____

____

6. It is possible to convert one form to another?

____

____

____

7. Do you agree or disagree with this quote? Why or why not?

____

____

____

"What is there that is not a poison? All things are poison and nothing (is) without poison. Solely the dose determines that a thing is not a poison." Paracelsus (1493-1541)

Now that you are completely fascinated by this topic, you decide to further investigate the chemistry involved in this lawsuit and discover that the hexavalent chromium had contaminated the aquifer that supplied the residents of Hinkley, CA. In addition, you find that PG&E told the residents in the vicinity of the plant to…

"….avoid drinking your well water, but it is safe to use for all other domestic purposes such as bathing and watering animals and plants." [Quote taken from the Anderson v PG&E case – Superior Court for the County of San Bernardino, Barstow Division, File BCV 00300].

8. How would you have responded to this suggestion?

____

____

This statement reminds you of a scene in the movie where Erin Brockovich has a discussion with some of the plaintiffs about the safety of swimming in their local pools (possibly contaminated with hexavalent chromium). You look some more into your research and discover that when the PG&E representatives were asked by residents about the safety of swimming in their pools, the PG&E response was basically that it was OK. In fact, the PG&E representatives went on to say that it was OK because the chlorine and other pool chemicals would "kill any contaminants in the pool, including chromium."

9. Is there any valid chemistry to the statement that chlorine and other pool chemicals will "kill any contaminants in the pool, including chromium"?

CASE 7: Consuming Chromium

Purpose: The purpose of this case study is to understand the molecular structure of chromium picolinate and its role as a supplement.

> After staying up very late at night trying to understand the chemistry involved in the "Anderson v Pacific Gas and Electric" lawsuit, you decide you may need some vitamins to boost your tired body (and brain). Once you're in the vitamin store you discover shelves with, none other than, chromium supplements claiming to "burn fat". Since you are now a self-proclaimed expert on chromium, you decide to examine the contents of these "fat-burning" supplements. They all contain chromium picolinate.

1. What is the structure of chromium picolinate?

____

2. What is the molecular weight of chromium picolinate?

____

3. What kinds of bonds are present in this molecule?

____

4. Now that you know there are two common kinds of chromium, what kind of chromium is present in chromium picolinate?

____

5. Are the claims that chromium picolinate helps to "burn fat" valid?

____

6. Are there any risks associated with ingestion of chromium picolinate?

____

7. What is the recommended intake of this supplement in order to "burn fat"?

____

8. Are supplements regulated by the U.S. Food and Drug Administration?

____

9. How does this impact potential risk from taking supplements?

____

Chapter 4: An Introduction to Organic Compounds

Related Learning Objectives

Define "electronegativity" and explain its relationship to polar covalent bonds. Give a simple rule that can be used to predict whether or not a covalent bond is polar.

List the five basic shapes about an atom in a molecule and describe the rules used to predict shape. Explain how shape plays a role in determining overall polarity.

Describe the four families of hydrocarbons.

Explain the difference between constitutional isomers, conformations, and the stereoisomers known as geometric isomers. Give examples of two different families of hydrocarbons that can exist as geometric isomers.

Define the term "functional group" and describe the features that distinguish hydrocarbons, alcohols, carboxylic acids, and esters from one another.

CASE 8: Molecules in the Movies

Purpose: The purpose of this case study is to understand the molecular structure of TCE and its role as an environmental contaminant.

"I'm a huge fan of John Travolta!" exclaimed Sally. "In fact, this weekend I'm hosting a 'John Travolta' film festival at my house and you're all invited. Each of you can pick your favorite John Travolta movie to bring."

"I'm bringing *Grease*"

"I'm bringing *Saturday Night Fever*"

"I'm bringing *Pulp Fiction*"

"I'm bringing *A Civil Action*"

"What?" everyone asked, "We never heard of *A Civil Action*. What's it about?"

"Well, it's about a lawyer, Jan Schlichtmann, played by John Travolta, who represents the people of Woburn, Massachusetts who claim two local companies contaminated their drinking water with a chemical called TCE and caused their children to contract and die from leukemia."

"What's TCE?"

1. Draw the structure of TCE.

2. Are there *cis* and *trans* isomers of TCE?

   ____

3. What was the TCE in the film "A Civil Action" used for?

   ____

   ____

   ____

4. How did the TCE become a contaminant in the water?

   ____

   ____

   ____

5. Should TCE be soluble in water? Why or why not?

____

____

____

6. What was the final verdict of this case?

____

____

____

7. Based on the chemistry that you've discovered, do you agree or disagree with this verdict?

____

____

____

____

____

CASE 9: Food and Forensics

Purpose: The purpose of this case study is to identify the polarity of compounds in order to utilize this information as a tool for multiple types of analysis.

To most of us, it looks the same, smells the same and tastes the same. So, what is the difference between pure vanilla extract and imitation vanilla? Well, for one thing, the cost. A four-ounce bottle of pure vanilla is about twice as expensive as the same amount of imitation vanilla. And, since vanilla is one of the most commonly used flavorings with market values in the millions of dollars per year, the food industry keeps a careful watch for imitation vanilla being sold as pure vanilla extract.

One way to monitor the purity of pure vanilla extracts is by a technique called gas chromatography. Gas chromatography (GC) is used to separate components of a mixture on the basis of molecular polarity. For the most part, the more polar components come through the instrument first followed by the less polar components. For pure vanilla extract, there are over 200 organic molecules present in the sample that are detectable with gas chromatography. Of these 200, only four of them are present in significant amounts - vanillin, vanillic acid, 4-hydroxybenzaldehyde, and 3,5-dihydroxybenzoic acid. In imitation vanilla, one of these components is missing and a less polar component appears in large amounts.

1. Draw the structure of the four major components of pure vanilla extract.

2. Arrange these predominant constituents of pure vanilla extract in order of decreasing polarity (what you'd expect to see on a gas chromatograph).

3. What would you expect the gas chromatogram of imitation vanilla to look like?

Now that you've developed an understanding of how to interpret and predict data generated from gas chromatography, let's put your new knowledge to work.

JN comes back to his fraternity house to find his roommate YP on the floor passed out. Knowing full well YP was attending a sorority party JN was concerned about his narcotic of choice for the evening. JN was unable to rouse YP and became increasingly alarmed. JN called 911 and they took YP to the local ER due to a suspected drug overdose. JN explained to the trauma team about YP's party habits and his agents of choice: barbiturates, opiates, THC, and his prescription use of Prozac®. In order to treat YP they decided to do a toxicology screen on his urine. The urinalysis was sent to the forensic chemist on staff who analyzed the sample by GC. The forensic chemist made up a standard sample containing a mixture of all the suspected agents. The standard mixture produced a chromatogram with elution times of 2 minutes, 4 minutes, 8 minutes, and 10 minutes. Then YP's urine sample was analyzed by GC. The chromatogram from YP's urine showed only one peak with an elution time of 4 minutes.

4. Match each drug to an elution time.

____

____

5. Explain your matches (question 1) based on the structure of the molecules in the standard samples.

____

____

____

6. From this GC data, determine the cause of YP's overdose.

____

____

____

____

Chapter 5: Gases, Liquids, and Solids

Related Learning Objectives

Convert between common pressure units.

List the variables that describe the condition of a gas and give the equations for the various gas laws.

Describe the relationship between atmospheric pressure and the boiling of a liquid.

CASE 10: Diving and Flying

Purpose: The purpose of this case study is to understand the physiological effects of extreme pressures.

"I'm so excited for our trip to the Caribbean!" exclaimed Sue. "We get on the airplane on Friday, do some sightseeing on Saturday, go scuba diving on Sunday afternoon, and then get on the plane to come home on Sunday evening."

"Uh oh" sighed John. "Are you sure our flight is on Sunday evening?"

"Yes" said Sue "I planned it that way so that we could maximize our fun and still get back to work on Monday."

"Well, that's not going to work. We won't be able to go scuba diving," said John.

"Why not?" asked Sue.

1. Why can't John and Sue scuba dive and fly in an airplane on the same day?

____

____

____

____

____

2. What is the chemistry behind this scenario as it relates to the gas laws?

____

____

____

____

____

CASE 11: Vomiting on Vacation

Purpose: The purpose of this case study is to understand various factors that affect the equilibrium of the liquid and gas phases.

EG, a 13-year-old teenager from Redondo Beach, California and his parents were vacationing in the French Alps atop the famous Col de la Madeleine. Wanting to do his part so his parents could have a relaxing vacation, EG decided to make them breakfast. Using his familiar techniques from home he made them each two slices of toast with butter, two eggs he put in boiling water and let steep for four minutes, a half cup of cottage cheese topped with three pear slices. Later that evening, after dinner, his parents had a fever, headache, abdominal cramps, nausea, vomiting, and cramps. The next morning they were also experiencing bloody diarrhea and were ill enough to visit the local emergency room.

1. What was EG's parents' bacterial diagnosis?

____

____

____

____

____

____

2. From a chemistry standpoint, how did EG's parents become infected with this bacterium?

____

____

____

____

____

____

Chapter 6: Reactions

Related Learning Objectives

Interpret and balance chemical equations.

Identify oxidation, reduction, combustion, and hydrogenation reactions.

Identify hydrolysis, hydration, and dehydration reactions of organic compounds.

CASE 12: All "-Caines" are not the Same

Purpose: The purpose of this case study is to identify the products of a hydrolysis reaction and to consider the biological consequences as they relate to humans.

NS, a 52 year old female, was being prepped for surgery. Upon initial HX and screening one of the questions asked was "Are you allergic to any foods or medications?" Being an astute chemist and thinking through the surgery to be performed, her original answer was "no." Upon further dialogue with the anesthesiologist, she finally remarked that she was allergic to Novocain®, even though she thought it impertinent to the procedure she was about to undergo because she was being given a general anesthetic vs. a local anesthetic. No further discussion took place as she was about to be wheeled off to the OR.

After surgery, upon waking from her anesthesia induced state she heard the anesthesiologist speaking to rouse her consciousness even further. During the bantering the issue of her previous allergic reaction to Novocain® came up. As they were discussing the chemistry of this allergic reaction, the nurse interjected, "How do you know what local anesthetic you can use on someone like her without doing a skin patch test to see what other local anesthetics she may be allergic to?" As the anesthesiologist was walking away he mumbled, "Anything with an 'i' before '-caine'."

As the nurse was still bewildered and NS was only semi-alert, please explain to the nurse the following:

1. What did the anesthesiologist mean by "Anything with an 'i' before '-caine'?"

____

____

____

____

2. What is the most likely chemical cause of NS's allergic reaction?

____

____

____

____

3. What is the primary difference between the class of compounds NS is most likely to be allergic to and the class of compounds she's not as likely to have an immunologic response to?

____

____

____

____

____

____

4. Show the reaction (using molecular structures) that takes place to produce the culprit of this allergic response.

5. What type of reaction have you represented in question 4?

____

____

____

____

CASE 13: Malaria and Methemoglobin

Purpose: The purpose of this case study is to comprehend the chemical reaction by which methemoglobin is formed.

The motto for the U.S. Army Research and Development Command is "Research for the Soldier." This encompasses a span of projects to include research on developing drugs to protect our soldiers from disease to developing agents to protect our soldiers from chemical warfare agents such as hydrogen cyanide.

One disease of concern is malaria due to the fact that our soldiers are stationed in areas of the world where malaria is prevalent. The Walter Reed Army Institute of Research (WRAIR) has been involved in developing anti-malarial drugs and a malaria vaccine for years. In the 1980s, WRAIR scientists discovered that a class of drugs known as the 8-aminoquinolines was potent against the malaria parasite. These drugs were about to be put into human clinical trials when a concerning side effect was noticed – they were promoting the formation of methemoglobin at significant levels. While the WRAIR scientists realized this side effect could be serious enough to "table" the 8-aminoquinolines for use as anti-malarials, they realized that they could take advantage of this side effect and that these compounds could be advantageous for another use related to their mission.

1. What is methemoglobin?

____

____

2. What kind of reaction occurs when an 8-aminoquinoline reacts in the body to form methemoglobin?

____

3. What is an alternative use (related to the mission of WRAIR) for these 8-aminoquinolines that takes advantage of the fact that they produce methemoglobin?

____

____

4. Describe the chemistry of this alternative use.

____

____

____

Chapter 7: Solutions, Colloids, and Suspensions

Related Learning Objectives

Define the term "pure substance" and describe what is meant by the terms "homogeneous mixture" and "heterogeneous mixture." For a homogeneous mixture, explain what differentiates solutes from a solvent.

Describe the effect that temperature has on the solubility of gases, liquids, and solids in water.

Use solubility rules to predict whether or not a reaction will produce a precipitate.

Explain the terms "hydrophilic," "hydrophobic," and "amphipathic" and give examples of compounds that belong to each category.

List some of the commonly encountered concentration units and perform calculations involving concentration.

Be able to calculate the concentration of a solution and the new concentration after it has been diluted.

List the differences between solutions, suspensions, and colloids.

Describe the principles of diffusion and osmosis.

CASE 14: Surviving a Ship-Wreck

Purpose: The purposes of this case study are to calculate various concentrations of solutions and to determine how to mix solutions so the resulting new solution is appropriate for drinking water.

You have unexpectedly found yourself stranded on a tropical island. After a thorough reconnaissance of the island, you realize that you have all the necessities for survival with one exception - fresh water. However, you did discover four wooden crates buried in the sand. You open the crates to see that each one contains a case of IV bags. Each case contains 24 x 1 L bags.

The contents of case 1 are labeled **D5W**.

The contents of case 2 are labeled **2/3 & 1/3**.

The contents of case 3 are labeled **half-normal saline**.

The contents of case 4 are labeled **saline**.

Upon examination of the bags, you find that they have not been tampered with and are not yet expired.

1. How can you maximize the use of these bags to enhance your survival?

___

___

___

___

___

___

___

___

___

___

___

___

Case 15: Vanishing Varicose Veins

Purpose: The purpose of this case study is to investigate different chemical mechanisms that ultimately yield the same physiological effect.

JJ, a 65-year-old woman, is disturbed by her telangiectasias, venules, and reticular veins. She went to a dermatologist who recommended the TX of sclerotherapy. The dermatologist explained that they would use an injected sclerosant to eliminate the abnormalities. Upon the day of treatment, JJ was interested in knowing about the injection solution and the doctor informed her it was a hypertonic saline solution at 23.4%. After the procedure JJ was told to wear Class I compression stockings at all times for two weeks, but that it may take up to three months for the discoloration to disappear.

Three months later JJ was disappointed to learn the TX didn't work at all. She consulted with the dermatologist who said they could try again with the same sclerosant; JJ decided against it.

A year later, JJ heard about laser treatments that could rid her of her affliction, only this time she consulted a vascular physician. During consultation the physician informed JJ that he could help her by sclerotherapy.

JJ was mortified saying "It didn't work at all last time. I want you to use the laser."

The physician asked her what sclerosant had been used last time and JJ said, "A hypertonic saline solution."

The doctor replied, "I'm going to use sodium tetradecyl sulphate instead. It works on a different chemical principle and I think this procedure will work for you."

JJ agreed to have the procedure using the sodium tetradecyl sulphate and was again told to wear Class I compression stockings at all times for two weeks. She was also reminded that it may take up to three months for the discoloration to disappear.

This time the procedure worked.

The doctor had commented that the two solutions worked on different chemical principles.

1. What are telangiectasias, venules, and reticular veins?

2. Explain the chemistry behind these two solutions and how each procedure has the potential to eliminate telangiectasias, venules, and reticular veins.

Case 16: Ineffective Antibiotics

Purpose: The purpose of this case study is to recognize reactions within a biological organism that results in poor bioavailability with potentially deleterious implications.

BA, a 72 year old female, presented at her NP's office with physical symptoms of dysuria, urinary frequency, pressure in the lower abdomen, pyuria, and a mild fever. Her NP asked for a urine sample but in the meantime decided to start BA on antibiotics because she suspected a UTI. BA was prescribed Tetracycline: 500 mg PO QID for 10 days.

BA went home and started her antibiotic regimen along with the following medications and supplements she takes:

Time	Medication/Supplement	Dosage	Time	Medication/Supplement	Dosage
0700	Glyburide	5 mg	**1700**	Glyburide	5 mg
	Mylanta® Ultra Tabs	2 Tablets		Mylanta® Ultra Tabs	2 Tablets
	Aspirin	80 mg		Atenolol	50 mg
	Atenolol	50 mg		HCTZ	12.5 mg
	HCTZ	12.5 mg		Tetracycline	500 mg
	Tetracycline	500 mg	**2200**	Tetracycline	500 mg
1200	Metformin	500 mg		Ambien ®	5 mg
	Mylanta® Ultra Tabs	2 Tablets			
	Tetracycline	500 mg			

After five days on this regimen and only minimal improvement of her symptoms, the NP changed RJ's prescription to Ciprofloxacin: 500 mg PO BID for 10 days. BA went home and started her antibiotic regimen along with the following medications and supplements she takes:

Time	Medication/Supplement	Dosage	Time	Medication/Supplement	Dosage
0700	Glyburide	5 mg	**1700**	Glyburide	5 mg
	Mylanta® Ultra Tabs	2 Tablets		Mylanta® Ultra Tabs	2 Tablets
	Aspirin	80 mg		Atenolol	50 mg
	Atenolol	50 mg		HCTZ	12.5 mg
	HCTZ	12.5 mg		Ciprofloxacin	500 mg
	Ciprofloxacin	500 mg	**2200**	Ambien ®	5 mg
1200	Metformin	500 mg			
	Mylanta® Ultra Tabs	2 Tablets			

After five days on this regime, there was no real improvement of her symptoms. The NP checked the sensitivity test run from BA's urinalysis and couldn't understand why Tetracycline or Cipro didn't eradicate the infection because the bacteria showed sensitivity to both of these antibiotics. The NP gave BA yet another antibiotic the bacteria showed sensitivity to: SMZ/TMP DS: PO, q12h for 10 days. BA went home and started her antibiotic regimen along with the following medications and supplements she takes:

Time	Medication/Supplement	Dosage	Time	Medication/Supplement	Dosage
0700	Glyburide	5 mg	**1700**	Glyburide	5 mg
	Mylanta® Ultra Tabs	2 Tablets		Mylanta® Ultra Tabs	2 Tablets
	Aspirin	80 mg		Atenolol	50 mg
	Atenolol	50 mg		HCTZ	12.5 mg
	HCTZ	12.5 mg		SMZ/TMP DS	1 Tablet
	SMZ/TMP DS	1 Tablet	**2200**	Ambien ®	5 mg
1200	Metformin	500 mg			
	Mylanta® Ultra Tabs	2 Tablets			

After 10 days on this antibiotic BA's UTI was gone.

1. Define the following: NP, dysuria, pyuria, urinalysis, antibiotic sensitivity, QID, BID, q.

____

____

____

____

____

____

____

____

____

____

2. Why did the NP think the infection was bacterial and not viral?

____

____

____

____

____

3. What is the primary function of each of the medications/supplements BA takes?

4. Why does SMZ/TMP DS not have a strength attached to the dosage?

5. Why did the tetracycline and ciprofloxacin not work for BA's condition even though the sensitivity test indicated they should have both worked on the bacteria?

6. Why did the SMZ/TMP DS work on the bacteria?

Chapter 8: Lipids and Membranes

Related Learning Objectives

Identify the basic steroid structure and list important members of this class of lipids.

Name the three types of eicosanoids and describe their biological function.

Describe the make up of a cell membrane and explain how various compounds cross the membrane.

Case 17: OTCs for Arthritis

Purpose: The purpose of this case study is to examine the mechanism of action of Vioxx® and to categorize over-the-counter medications for arthritis according to their mechanism of action.

In January 2004, EU, a 71-year old Vietnam veteran, was telling his daughter-in-law (a medicinal chemist) that his physician at the VA hospital had him on a prescription of Vioxx® for his arthritic shoulder. The daughter-in-law had a fit and told EU to stop taking the Vioxx® immediately. She was well aware of the deleterious side-effects of Vioxx® and other COX-2 inhibitors and assured EU that Vioxx® would soon be taken off the market. EU then asked his daughter-in-law about alternatives for Vioxx®. He was still active as a lumber farmer and wanted to maintain his lifestyle. She recommended, for now, that he should use an over-the-counter arthritis medication until he could get a new prescription. EU went to the pharmacy and found many over-the-counter medications for arthritis including: Tylenol®, Advil®, Motrin®, Aleve®, Anacin®, Bayer®, and Bufferin®.

1. Make a recommendation to EU for one or more of these OTC medications. Support your recommendation with appropriate biochemistry and assumptions if necessary.

____

____

____

____

____

____

____

____

2. Which of these are most like Vioxx®? Why?

____

____

____

____

____

____

Case 18: Patch Me Up!

Purpose: The purpose of this case study is to analyze structural features of medications that make them suitable for administration by a patch formulation.

> DC recently visited his local pharmacy to ask about medications for motion sickness. He is going on a cruise for his honeymoon and is worried about getting sick and ruining their trip. His pharmacist recommended a patch formulation for a motion-sickness drug. DC was not familiar with this form of medication and asked the pharmacist more about these "patches". "Oh, they're great," remarked the pharmacist. "It's too bad they can't be used for all medications – that would save us a lot of worry and money. Right now, they can only be used for hormonal treatments, to help stop smoking (nicotine patches), for motion sickness, to help with anxiety, to control an overactive bladder, and to deliver nitroglycerin to patients who've suffered a heart attack."

1. Why does the pharmacist think "they're great"? Discuss the advantages and disadvantages of using a patch formulation.

2. Why does a patch "save us a lot of worry and money"?

3. Discuss the medications that are delivered in a patch formulation. How are they similar? What properties do they share that allow them to be used in a patch?

4. Why are patches limited in their use?

Chapter 9: Acids, Bases, and Equilibrium

Related Learning Objectives

Define Bronsted-Lowry acids and bases and explain how they differ from their conjugates. Relate acid strength to conjugate base strength.

Explain how the pH of a solution can affect the relative concentration of an acid and its conjugate base and describe buffers.

Case 19: Only Half Awake

Purpose: The purpose of this case study is to decipher properties of acid-base chemistry and how they adhere to Le Châtelier's principle.

> YZ, a 55 year old male, was up at 5 AM preparing for his day at work. YZ suffers from keratoconjunctivitis sicca and uses Restasis® on a regular basis. Unfortunately, he hadn't put his contact lenses in for the day and when reaching into the medicine cabinet he grabbed his *Volsol HC* otic drops and put those in his eyes instead. The pain was excruciating and he was taken to the local ER by his wife. The ER LPN flushed his eyes with a saline solution drip for one hour. YZ's eyes felt much better and he was discharged with the recommendation to use *Collyrium for Fresh Eyes* if there was continued discomfort. As YZ was reluctant to put anything in his eyes he read the labels of all of these products. He was appalled when he saw an acid as one of the ingredients in the optic drops his discharge papers had suggested. He called the ER to complain about their erroneous product recommendation. You are the pharmacist on duty and the call is referred to you.

1. What are the primary uses of the three products mentioned?

____

____

____

____

____

____

____

____

2. What are their active ingredients?

____

____

____

____

____

3. Why did the acid from the otic drops hurt his eyes but not the acid in the *Collyrium for Fresh Eyes*?

____

____

____

____

____

4. Explain to YZ the purpose of the acid in each of these products.

____

____

____

____

____

____

____

5. Your supervisor overhears your conversation with YZ and would like you to draw all of the acid-base reactions taking place in this scenario. He wants you to explain them to your pharmacy technicians since they are capable of understanding the chemistry involved in this case.

____

____

____

____

____

____

Case 20: Diaper Rash Dilemma

Purpose: The purpose of this case study is to understand the acid-base chemistry associated with diaper rash and how this chemistry pertains to a possible treatment for diaper rash.

> RL, a 28-year old first time father, was watching his 4-month old baby for the weekend while his wife was away at a sales meeting. Things were going well. He had run out of breast-milk to give the baby, so he went to the store to get some formula. The baby responded well to the formula - consuming an entire bottle in one feeding. In an hour or so he noticed a smell and figured it was time to change the baby's diaper. To RL's horror, the baby had a very wet diaper (to include diarrhea) and a bright red rash all over her butt. After he cleaned up the mess, he looked around the baby's changing table for remedies for this horrible rash, but he couldn't find anything. As he was about to panic, he remembered some useful acid-base chemistry from a class years ago. He opened the refrigerator and found something that would help alleviate the rash.

1. What could've AJ used to help with the diaper rash? From a chemical standpoint, how does it work?

____

____

____

____

____

2. What is the acid-base chemistry of this case?

____

____

____

____

____

3. Write the balanced chemical equation(s) that represent this chemistry.

Chapter 10: Carboxylic Acids, Phenols, and Amines

Related Learning Objectives

Describe what takes place when carboxylic acids and phenols react with water or with strong bases.

Describe what takes place when amines react with water or with strong acids.

Explain how carboxylic acids can be converted into esters and amides.

Explain the terms chiral molecule and chiral carbon atom, distinguish between enantiomers and diastereomers, and define the terms dextrorotatory and levorotatory.

Case 21: Treatments for Toothaches

Purpose: The purpose of this case study is to look at the acid-base chemistry of a potential drug-drug interaction.

> A 4-year old boy is brought by his mother into the emergency room on a Friday night. The boy is complaining of severe tooth pain (he's actually screaming "OW! OW!"). The attending physician takes a look in his mouth and notices substantial swelling in his gums and cheeks and the appearance of an abscessed tooth. The physician asks his mother some medical questions and discovers he's a healthy 4-year old with no history of allergies. The mother has been giving her son children's ibuprofen in a suspension formula (and she's been taking ibuprofen tablets herself for the stress!). She has also been putting OraGel® on the painful area. The physician prescribes an antibiotic to treat the infection (the abscessed tooth) and recommends to the mother to continue giving her son ibuprofen to help with the pain and swelling. As an ER physician, he's in a hurry, so he starts to dash out the door when the mother asks, "Is it alright to give the antibiotics and ibuprofen at the same time? He hates taking medications, so if I can give it to him all at once, it would be much easier." But it was too late – the physician was already gone.

1. Draw possible chemical reactions that could occur between the ibuprofen and the antibiotic (you have to choose one or more antibiotics).

2. Based on your chemical analysis above, is it OK to give the boy the ibuprofen and the antibiotic at the same time? Why or why not?

____

____

____

____

____

On Monday morning, the mother is able to get an appointment for her son at the pediatric dentist in town. The dentist takes one look in the mouth and states that oral surgery will be necessary. He will have to sedate the child in order to perform the surgery. The dentist then asks the mother if her son is currently taking any medications. She shows him the prescription for the antibiotic and states that he's also been on a continuous regime of ibuprofen. The dentist sighs and says, "We may need to hold off on the surgery for a day or so." The mother, who is at wits end and can't stand to wait another day, demands to know why the surgery has to be delayed.

3. Why does the surgery need to be delayed?

Case 22: pH and the Parasite

Purpose: The purpose of this case study is to prompt the student to read an article from the primary literature and to interpret and extrapolate the data presented in this article for potential clinical uses.

Go to the following website:

http://www.pubmedcentral.nih.gov/articlerender.fcgi?artid=397650

AVNER YAYON, Z. IOAV CABANTCHIK, AND HAGAI GINSBURG
Proc Natl Acad Sci U S A. 1985 May; 82(9): 2784–2788.
"Susceptibility of human malaria parasites to chloroquine is pH dependent."

1. Draw the structure of chloroquine at an acidic, neutral, and basic pH.

2. How does the change in pH affect the ability of the chloroquine to permeate a cell membrane?

____

____

____

____

____

3. In what ways can the research described in this article impact the clinical use(s) of chloroquine?

____

____

____

____

____

Case 23: Which Vitamin C is Best for Me?

Purpose: The purpose of this case study is to examine the structures of two forms of vitamin C and how their structure relates to their use.

> You recently came across an advertisement in a wellness magazine for a new form of vitamin C that is "more bioavailable". The ad claims this vitamin C will work better and last longer. Luckily, you are in a great chemistry class that integrates organic and biochemistry into the curriculum, allowing you to understand the word "ester" present in the ad. A friend stops by while you're reading the magazine and states, "Wow. That sounds great. I'm going to buy that new form of vitamin C." You respond that she'd be wasting her money, unless of course, she would want to use it as part of a skin cream.

1. What chemistry did you have insight to that allowed you to respond in this way?

Case 24: Medicine in the Mirror

Purpose: The purpose of this case study is to compare medications that are distributed as pure enantiomers or racemic mixtures.

Now that your friends have realized that you are really learning a lot of interesting chemistry in this integrated format, they are constantly asking you questions. The latest question that has arisen comes from an email circulating within the pharmacy in a local hospital. The debate that is being conducted via email concerns the use of Prevacid® vs. Nexium®. While both drugs have the same therapeutic effect, one is sold as something called a "racemate" and the other is sold as something called a "single enantiomer." The email debate specifically addresses the issue of the safety of the one sold as a racemate. Some of the pharmacists assert that it's safe, some say it's not. One of the pharmacists brings up data associated with the drug Advair® to support his claim.

1. Discuss the safety of medications sold as a racemate.

____

____

____

____

2. Discuss the chemistry (and safety) of the drugs Prevacid® and Nexium® with respect to the terms "racemate" and "single enantiomer."

____

____

____

____

3. Why did a pharmacist bring up Advair® – how is that drug related to this discussion?

____

____

____

____

Chapter 11: Alcohols, Ethers, Aldehydes, and Ketones

Related Learning Objectives

Describe the structure of molecules that belong to the alcohol and ether families.

Distinguish 1°, 2°, and 3° alcohols.

Describe the oxidation reactions of alcohols.

Case 25: Sweet, Clear, Colorless and Deadly

Purpose: The purpose of this case study is to understand the biochemical oxidation of an alcohol and its lethal consequences.

> MO and his roommates recently returned from a winter ski-trip. As they walked into the house, MO was horrified to see his 2-year old dog was lying unconscious. Upon closer inspection, he sadly realized his dog had died when they were away. The dog was previously very healthy, was kept in a house with a 'doggie door' to the garage, and was well cared for. This death was a complete mystery to MO. He immediately called an emergency veterinary clinic and the on-call vet asked MO a series of questions to include an inquiry about what's stored in their garage. MO told the vet the dog had access to the garage and that they had the usual stuff in there - tools, lawn mower, a bucket of anti-freeze, bikes, and a wheel barrow. The vet told MO to bring the dog in for pathology, but he was pretty sure he knew the cause of death already.

1. What does the vet suspect to be the cause of death?

____

____

____

2. Describe the biochemical processes involved in this toxicity.

____

____

____

____

____

____

3. What could've been done to counteract this toxicity?

____

____

____

____

____

Case 26: Gallstones and Gasoline

Purpose: The purpose of this case study is to link an epidemiological trend to pertinent environmental chemistry of an area.

You have just been hired as an epidemiological nurse to investigate the problems of obesity and high cholesterol in Native American populations. Your first task is to gather data on the Navajo tribe in northern Arizona. Upon arrival, you are shocked to discover that this is a cold climate due to the high elevation (about 7,000 ft. above sea level). You trudge through the snow to the local clinic to begin reviewing the charts of the patients. One indicator of high cholesterol is the incidences of gallstones. Typically, Native American women have an 80% chance of developing gallstones during their lives, yet upon inspection of the data you have gathered, only 10% of the Navajo women have developed gallstones. You are puzzled by this difference considering that the other data on the Navajo women shows 60% of the women are obese and their fasting serum triglyceride concentrations were high (median: women, 137 mg/dL), and concentrations of HDL cholesterol were low (median: women, 44 mg/dL).

Later in your visit, you are driving through the Navajo Reservation and notice several abandoned gasoline stations. You have a "light bulb" moment and decide to make a quick phone call to the Navajo Nation Environmental Protection Agency about these abandoned gas stations and their water quality issues. You discover that these gas stations were built in the 1970s and closed in the late 1990s. They have several underground storage tanks.

1. What is the connection between the low incidences of gallstones and the abandoned gasoline stations?

2. Why did you want to know about the water quality in this region? What piece of information were you looking for?

____

____

____

____

____

____

____

____

____

3. What is your hypothesis concerning the statistically lower incidences of Navajo women with gallstones?

____

____

____

____

____

____

____

____

Chapter 12: Carbohydrates

Related Learning Objectives

- Describe the difference between mono-, oligo-, and polysaccharides and explain the classification system used to categorize monosaccharides.
- Identify the four common types of monosaccharide derivatives.

Case 27: Scanning for Sugars

Purpose: The purpose of this case study is to be aware of carbohydrates within the context of biological functionality.

SL, a 28 year old male, was diagnosed with lung cancer. In order to ascertain if the cancer had spread to adjacent lymph nodes, an 18-FDG PET scan was ordered. SL meticulously followed the prescribed preparation for the exam. On the day of the exam, perfunctory blood work was also drawn. When the scan was completed the results were inconclusive because there was too much "background activity" throughout his entire torso.

Two days later SL's blood results were sent to his physician's office where an astute nurse took a quick glance over them.

Test	**Result**
RBC	6.0 million/mm^3
WBC	9,500/mm^3
Hematocrit	45.6%
Hemoglobin	16.6 g/dL
MCV	92 femtoliter
MCH	30 pg/cell
MCHC	35 pg/cell
RBS	14 mmol/L
Thrombocyte	400,000/ mm^3

From the results, she immediately recognized the problem that resulted in the inconclusive exam. After consulting with the oncologist, she reordered the scan.

1. What is an 18-FDG PET scan?

____

____

____

2. How does it work?

____

____

____

____

3. What do each of these blood test acronyms stand for? What are the normal values for each test?

____

____

____

____

____

____

____

4. What is it the nurse saw that helped her determine the scan could be reordered with success this time?

____

____

____

____

____

____

____

5. What is it the medical staff will have to do this time with the patient to help insure the success of the 18-FDG PET scan?

____

____

____

____

____

____

____

Case 28: A Polyhydroxy What?

Purpose: The purpose of this case study is to assess the structural components, variability, and multiple functionalities of carbohydrates.

You have just received a prestigious position as an undergraduate research assistant, at a prominent research institution, for the summer. Your mentor, EM, is a graduate research assistant in the biological chemistry department. EM shows you around the department at 8 AM and informs you that your responsibility in the program is being the resident expert on carbohydrates for HIV.

You explain to EM, "I took a biochemistry course last semester but I didn't really understand carbohydrates. In fact, I got a "D" on that exam even though I pulled a "B" in the course."

EM replies, "Well, you must have fabulous research and analytical skills or you would have never received this position. We had 50 applicants for this one position YOU were given."

You inquire of EM, "Could you help me bone-up on my carbohydrate chemistry?"

EM points in the direction of a tall, six story building, "There's the library, inclusive of our extensive science library with plenty of sophisticated data bases. I will see you back here at 6 PM tonight. I expect you will be conversant and knowledgeable about every facet of information possible in regards to carbohydrates and their relationship to HIV. Welcome to the world of research! Au revoir."

You are left standing there alone and petrified that you have absolutely no working knowledge of carbohydrates. You stumble into the science library and obviously look bewildered. The librarian notices your pale and disoriented expression and asks if she can help. You explain your predicament and she provides the following guidance.

1. Recall what you've heard from the media; it's a virus we don't have many answers to especially in regards to a vaccine or a large impact on delaying its progression.

____

____

____

____

____

____

2. See if carbohydrates are associated with the virus. If they are, what role do they assume?

____

____

____

3. If carbohydrates play a role, do they help the virus or hurt the virus? Do they help the researchers or hurt the researchers with their mission?

____

____

____

____

4. What impact do carbohydrates have in regards to finding a vaccine?

____

____

____

____

5. What impact do carbohydrates have in regards to delaying progression of the disease?

____

____

____

____

6. Where does current research stand at this point in time? In other words, what fronts are researchers working on in order to find a vaccine or delay for disease progression especially in regards to carbohydrates since that is what you are responsible?

____

____

____

____

____

____

Chapter 13: Peptides, Proteins, and Enzymes

Related Learning Objectives

Describe the structure of amino acids and the system used to classify amino acids.

Distinguish between oligopeptides, polypeptides, and proteins and describe the bond that joins amino acid residues in these compounds.

In terms of peptides and proteins, define primary, secondary, tertiary, and quaternary structure. Name the covalent and noncovalent forces responsible for each level of structure.

Explain what is meant by the term denaturation, and list some of the ways to denature a protein.

Distinguish between absolute specificity, relative specificity, and stereospecificity.

Name the steps required for a typical Michaelis-Menten enzyme to convert a substrate into a product. Describe K_M and V_{max}, explain how competitive and noncompetitive inhibitors affect these two parameters, and distinguish between reversible and irreversible inhibitors.

Case 29: The Case of the Red Hot Peppers

Purpose: The purpose of this case study is to recognize the relationship between molecular structure and receptor site theory.

After eating a bowl of chili laden with hot chili peppers, PP exclaims, "Ouch. My mouth is on fire. Give me some water!"

"That won't work," says AH. "Try some milk instead."

"Why milk?" asks PP. "And, why aren't you in pain? You're eating that chili by the spoonful, like there's nothing to it."

"Oh, I eat hot foods all the time," says AH. "I guess I've gotten used to it or something."

1. What class of molecules makes chili peppers "hot"?

____

____

____

____

2. How do they fit into your receptors?

____

____

____

____

3. Why milk?

____

____

____

4. Why doesn't AH have the same sensation?

____

____

____

____

Case 30: A Boost for Red Blood Cells

Purpose: The purpose of this case study is to develop a hypothesis as to why a protein-based drug is not producing the expected results.

MK, a 51-year old male has been diagnosed with lung cancer and has been undergoing a series of chemotherapy treatments for six weeks. During this time, MK has become very tired and sometimes breathless from normal activities. After the last visit to his oncologist, he discovered that he has anemia and his physician prescribed a regime of Procrit® to combat the anemia and tiredness. MK was instructed on the administration of Procrit® as it requires a subcutaneous injection under the skin. The pharmacist also made special note of the storage of this medication. One month later, MK is still feeling very tired and breathless -- he noticed no change from taking the Procrit. His oncologist is monitoring the results of his hematocrit and is frustrated by the lack of improvement.

1. What is Procrit®? How does it work?

____

____

____

____

____

2. Why is it necessary to give it subcutaneously?

____

____

____

____

3. What are the guidelines for storing Procrit®? Why?

____

____

____

____

____

4. What is your hypothesis for the lack of improvement in MK's hematocrit? Defend your answer with biochemical reasons.

5. How would you test your hypothesis?

Chapter 14: Nucleic Acids

Related Learning Objectives

- Describe the makeup of nucleosides, nucleotides, oligonucleotides, and polynucleotides.
- Describe the primary structure of DNA and RNA and secondary and tertiary structure of DNA.
- Explain how replication takes place and describe the roles of DNA polymerase in this process.
- Explain how transcription takes place and describe the role of RNA polymerase in this process.
- Name the three types of RNA and identify the role of each in translation.
- Explain the term mutation.

Case 31: Surviving with Sickle Cell Anemia

Purpose: The purpose of this case study is to understand the pros and cons of a genetic mutation.

1. Do you consider the genetic mutation resulting in sickle cell anemia (SCA) to be a "positive" or "negative" mutation? Why? Be sure to include a discussion of the mutation, how the mutation occurs, and the pros and/or cons of the mutation.

Chapter 15: Metabolism

Related Learning Objectives

Define the terms metabolism, catabolism, and anabolism.

Describe the three different types of metabolic pathways and explain what coupled reaction means.

Name the products formed during the digestion of polysaccharides, triglycerides, and proteins and state where the digestion of each takes place.

Identify the initial reactant and final products of glycolysis, describe how this pathway is controlled, and explain how gluconeogenesis differs from glycolysis. Describe how the manufacture and breakdown of glycogen are related to each of these pathways.

Give an overview of the citric acid cycle, explain how it is controlled, and describe how the products of this circular pathway are used by the electron transport chain and oxidative phosphorylation.

Describe the catabolism of triglycerides, the β oxidation spiral, and how the β oxidation spiral differs from fatty acid biosynthesis.

Explain the fate of the amino groups in amino acids.

Case 32: Intermingled Pathways

Purpose: The purpose of this study is to assimilate various metabolic pathways and the complexity of the dependency of their components.

A couple was the proud new parents of a baby girl they named CC. CC initially presented as a normal, healthy newborn. By day seven, CC's diapers were stained black where she had urinated. The couple immediately took CC to the pediatrician where she was diagnosed with alcaptonuria. The pediatrician referred them to a genetic counselor. During their meeting with the genetic counselor, the couple was distraught by thoughts that their daughter was in imminent danger and were overwhelmed with information they couldn't comprehend.

The couple returned home to share the information with their extended family. The understanding of the information they walked away with was relayed to the family by them, "We have no idea what the counselor was talking about. We're more frightened now than when we walked in." They added, "We were told CC has a metabolic disorder related to tyrosine but we distinctly remember ALL the metabolic pathways from our college courses and they were glycolysis, TCA cycle, β-oxidation, and the electron transport chain. We remember tyrosine was an amino acid but what does that have to do with metabolism?" Now trembling, they said, "The counselor also brought up PKU in the discussion. We are aware that is something they test for in newborns before they leave the hospital. The counselor claimed CC doesn't have PKU but we should have a broad understanding of how all these pieces in the pathway are related. But even at that we were then instructed to restrict CC's intake of phenylalanine. We just don't understand why we have to restrict her phenylalanine intake if she has a tyrosine disorder." One of them, practically in tears at this point, mumbled, "Then the counselor tried to show us how a lack of melanin causing albinism is related to tyrosine. I can't figure out if she's trying to tell us that CC also has this disease. Why would she even bring this up?"

This couple is quite typical of most new parents; concerned about their child who they were unaware would be born with an affliction they knew nothing about. With all the emotional turmoil going on, it is not uncommon for people to misunderstand or not understand anything that is being explained to them by the medical professional.

Since you are a practitioner who is knowledgeable in various aspects of science and medicine and more importantly you are empathetic to the needs of new parents, could you take the time to explain this mess in a way the couple will understand?

1. Draw out the pathway that shows how phenylalanine, tyrosine, and melanin are related.

2. Explain the disease aspects of PKU, albinism, and alcaptanuria being sure to point out in the pathway you've drawn where the deficit occurs.

3. Explain that the only metabolic disease CC has is alcaptanuria and what they can expect from this disease throughout her lifespan.

4. Using the pathway you've created, explain to them why they need to restrict both tyrosine and phenylalanine in her diet.

5. Lastly, let the couple know if there are any other medications or palliative treatments they can give CC to help with her disorder.

Case 33: WARNING! Detour Ahead

Purpose: The purpose of this study is to incorporate individual metabolic pathways into a cohesive unit and deduce the mechanistic regulators of their interdependency.

DK is a 77 yo male presenting at your emergency room via ambulance. He has had a massive MI and CPR is underway. DK has finally been resuscitated but the trauma team realizes he is quite unstable and has cardiac ischemia. Your hospital is involved in a FDA approved drug study that involves the administration of dichloroacetate in patients with massive MI's. The chief resident on duty believes DK is a great candidate for this study and approaches the hospital's Patient Advocate, who is a part of the drug study team at this hospital - YOU! The resident asks you to approach DK's family with full disclosure of the medication study and to obtain an informed consent form for the procedure.

Although you knew you were on the drug study team, this is the first time you've had to explain the situation to the family and obtain their informed consent; you are nervous. Being conscientious and wanting to be respectful of both the medical team and the family, you take a few minutes to gather your information and your thoughts because your only thought at this moment is that dichloroacetate involves three metabolic pathways: glycolysis, β-oxidation, and the TCA cycle. You've formulated the approach in your mind on how to present the information to this family in a sympathetic, supportive, and understandable way that families in crisis may find difficult to comprehend.

1. Draw a schematic diagram of how all three of the pathways converge, labeling their pertinent starting material and ending products. No need to overwhelm them with all the enzymes and regulators.

2. Draw the critical enzymes and regulators that dichloroacetate will induce. This will essentially be an enlarged schematic diagram of the regulation points of the three pathways induced by dichloroacetate.

3. Explain in terms the family will understand what dichloroacetate is hypothesized to accomplish in order to help their loved one.

4. Since it is a drug under investigation at this point you must let the family know some of the problems it could cause. You are well aware of drug testing procedures in this country and there have been animal studies done with this potential medication.

5. Because you attended the mandatory classes in order to be a part of the drug study team, you are familiar with the aspects of other debilitating diseases that would render this same medication appropriate for treatment, prevention, or use as a palliative for other diseases. List some of these diseases where it appears dichloroacetate will be of potential benefit.

NOTES

NOTES

NOTES

NOTES

NOTES

NOTES

NOTES

NOTES

NOTES

NOTES

NOTES

NOTES

NOTES

NOTES

NOTES

NOTES

NOTES

NOTES

NOTES

NOTES

NOTES